Amazing Nature

Fantastic Feeders

Tim Knight

www.heinemann.co.uk/library
Visit our website to find out more information about **Heinemann Library** books.

To order:
☎ Phone 44 (0) 1865 888066
📄 Send a fax to 44 (0) 1865 314091
💻 Visit the Heinemann Bookshop at www.heinemann.co.uk/library to browse our catalogue and order online.

First published in Great Britain by Heinemann Library, Halley Court, Jordan Hill, Oxford OX2 8EJ, part of Harcourt Education.
Heinemann is a registered trademark of Harcourt Education Ltd.

© Harcourt Education Ltd 2003
The moral right of the proprietor has been asserted.

All rights reserved. No part of this publication may be reproduced, stored in a retrieval system, or transmitted in any form or by any means, electronic, mechanical, photocopying, recording, or otherwise, without either the prior written permission of the publishers or a licence permitting restricted copying in the United Kingdom issued by the Copyright Licensing Agency Ltd, 90 Tottenham Court Road, London W1T 4LP (www.cla.co.uk).

Editorial: Jilly Attwood and Claire Throp
Design: David Poole and Geoff Ward
Illustrations: Geoff Ward
Picture Research: Peter Morris
Production: Séverine Ribierre

Originated by Ambassador Litho Ltd
Printed and bound in Hong Kong, China by South China Printing Company

ISBN 0 431 16650 1
07 06 05 04 03
10 9 8 7 6 5 4 3 2 1

British Library Cataloguing in Publication Data
Knight, Tim
Fantastic Feeders – (Amazing Nature)
591.513
A full catalogue record for this book is available from the British Library.

Acknowledgements
The Publishers would like to thank the following for permission to reproduce photographs:
Bruce Coleman pp. **11** (Kim Taylor), **16** (Gunter Ziesler), **19** (Kevin Cullimore), **21** (Christer Fredriksson), **24** (Jeff Foott); Corbis p. **27**; FLPA pp. **5**, **13**, **14** (Minden Pictures), **10bl** (R van Mostrand), **25** (Silvestris); NHPA pp. **4** (John Shaw), **8** (Kevin Schafer), **9** (Martin Harvey), **10tr**, **18** (Nigel J Dennis), **12br**, **17** (Stephen Dalton), **12cr**, **20** (Anthony Bannister), **15**, **22** (Stephen Krasemann), **23** (Daniel Heuclin); Tim Knight pp. **6**, **7**, **26**

Cover photograph of the Venus fly-trap, reproduced with permission of the Bruce Coleman Collection.

Every effort has been made to contact copyright holders of any material reproduced in this book. Any omissions will be rectified in subsequent printings if notice is given to the Publishers.

Contents

The giant restaurant	4
Plant food	6
Eat your greens	8
Plants bite back	10
A sweet tooth	12
Fruit and nut cases	14
Insect collectors	16
Helping each other	18
Blood suckers	20
Big appetites	22
Predators	24
Nature's vacuum cleaners	26
Fact file	28
Glossary	30
Index	32

Any words appearing in the text in bold, **like this**, are explained in the Glossary.

The giant restaurant

All over the planet plants and animals are eating or being eaten. Even as you read this book, an animal somewhere in the world is in the middle of a feeding frenzy. The natural world is like a giant restaurant that never closes.

Omnivores are animals that will eat almost anything. Brown bears, for example, will pick berries, eat dead animals, catch fish or steal sandwiches from a campsite. Other animals are very fussy eaters. Many caterpillars can only eat the leaves of one particular kind of plant. Without these leaves, they would starve to death. Most animals stick to the same kind of food every day. They are experts at finding, catching and eating it.

An unlucky migrating salmon ends up as a meal after leaping straight into the jaws of a hungry grizzly bear.

Giant spiders like this tarantula are powerful enough to catch and kill birds, lizards and frogs.

Struggle to survive

Although their eating habits may be totally different, all animals and plants face the same problem. They all have to find enough food to stay alive, grow bigger and **reproduce**. Each one has developed its own way of dealing with this problem. Giant anteaters feed on ants that live in underground nests. They poke their long, sticky tongue through the narrow entrance and use it to reach the ants deep inside. The egg-eating snake has expandable jaws, so it can swallow an egg bigger than its own head.

Plant food

Plants do not have mouths or stomachs, but they still have to feed. They collect water and **minerals** through their roots. They also use sunlight to make their own food. They take in sunlight through their leaves. This amazing process is called **photosynthesis**.

Different plants need different amounts of sunlight. Giant aroids grow in the Borneo rainforest, where hardly any sunlight reaches the dark forest floor. The plants survive by sprouting enormous leaves to catch as much sunlight as possible.

Some plants in the rainforest need lots of sunlight. They grow in the treetops where there is plenty of light. Their roots cannot reach the soil far below so they take in water from the damp air. These plants are called **epiphytes**. If pools of rainwater collect among the epiphyte's own leaves, tiny creatures make their home there. Their droppings and dead bodies rot too. This gives the plants an extra helping of **nutrients**.

A bird's nest fern growing high on the trunk of a small tree collects water and plant food among its leaves.

Killer plants

If the soil where they grow has no nutrients, some plants need to kill insects and other small animals for food. Pitcher plants have vase-shaped traps growing from their leaf-tips. These traps contain an **acidic** liquid. When insects crawl inside the traps to look for **nectar**, they drown. The plant uses their rotting bodies for food. The sundew's leaves are covered in sticky hairs. If an insect lands on this deadly glue, the leaf curls around it and starts to **dissolve** the insect's body.

The Venus fly-trap can close its leaves on an insect in less than a third of a second.

Eat your greens

Many animals, especially insects, are expert plant munchers. Some animals nibble a leaf or two, some suck out the plant's **sap**, others gobble the whole plant. While the world's plants are turning sunlight into food and growing new leaves, billions of hungry jaws are busy eating them. For an animal, every leaf is a miniature feast. Leaves are packed with sugar and **starch**, which the animal uses for energy.

Leaf-cutter ants bite off sections of leaf and carry them underground. They chew the leaves into a soggy pulp, and feed on the fungus that grows on the rotting leftovers.

Eating machines

When caterpillars gather in huge numbers and move about eating all that they can find, they can become serious **pests**. Plagues of gypsy moth caterpillars can strip the leaves from every tree in the neighbourhood, including valuable fruit trees. The desert locust is the world's most destructive insect. In a single day, a swarm of 50 million locusts can eat enough food to feed 500 people for a year.

An animal has to break food up into tiny pieces before its body can use it. This process is called digestion. The main part of a plant cell, **cellulose**, makes leaves tough and hard to digest. Most plant-eating animals rely on **bacteria**, living in their stomach, to do this job for them. Some animals, known as **ruminants**, chew their food more than once. Llamas, camels, goats and reindeer are ruminants. They have large stomachs divided into three or four separate parts to help them digest all the plants they eat.

The proboscis monkey, which eats huge quantities of mangrove leaves, has a large pot-belly to help it digest them.

Plants bite back

Many plants have developed ways to avoid being eaten. They grow out of reach or defend themselves with thorns, stings and poisons. Most animals leave these plants alone because they are so difficult to eat. Any animal that works out how to eat these types of plants will have a rich supply of food all to itself.

Eating a mouthful of grass is like chewing hundreds of tiny swords. Every blade of grass is as sharp as a razor. Ordinary teeth would quickly wear away if they were used for **grazing** day after day. Zebra, bison, kangaroos and other grazers have teeth with open roots. This means their teeth can keep growing until the animals die. The teeth are pushed upwards as their upper surface wears out, like the lead in a propelling pencil.

Some caterpillars can eat even the most poisonous plants without coming to any harm. Cinnabar caterpillars store the ragwort's poison in their own bodies, protecting themselves from hungry birds.

Giraffes use their long necks to reach the highest branches of Africa's tall acacia trees. Their mouths have a tough lining that protects them from the tree's prickly thorns.

Some plants have poisonous leaves, which most animals cannot eat. Leaf-eating monkeys choose only the freshest leaves, which are not as poisonous as the old ones.

The okapi was only discovered by scientists a hundred years ago in the rainforests of Central Africa.

Giraffe Links

The giraffe and it's shorter-necked cousin, the okapi, both have an incredibly long, powerful tongue. They use it to pluck bunches of leaves from high branches.

Giraffes have an extra large heart to pump blood all the way up their neck.

A sweet tooth

Some animals feed on **nectar**. This is a sweetened liquid produced by plants to attract insects and other sugar seekers.

Many creatures have special body parts to help them reach nectar. The South American sword-billed hummingbird uses a beak longer than its own body to reach deep inside flowers. The Australian honey possum has become so used to sipping nectar that its teeth have almost disappeared.

A feeding hawkmoth uses its proboscis like an extra-long straw to suck up nectar.

Hover Links

Hawkmoths and hummingbirds can both drink while hovering in mid-air. This allows them to feed from delicate flowers that cannot take their weight.

Hummingbirds lick up nectar by flicking their long tongues into a flower ten times a second.

Bloated honeypot ants hang from the roof of their underground nest chamber, waiting to be milked by the workers.

Living honeypots

In countries with a short flowering season, the nectar soon runs out. Honeypot ants have a way of storing it. Some of them become living jam jars. All summer, the worker ants gather nectar. They bring it back to their underground chamber and feed it to other ants in the colony, called repletes. Stuffed so full that they cannot walk, the repletes simply hang upside down, blown up like little golden balloons. When the flowers stop making nectar, the hungry workers 'milk' the nectar from the repletes. The worker ants refill them later.

Fruit and nut cases

Some animals are much more interested in the fruit and seeds that plants produce after they have flowered. A fruit's flesh is juicy and tasty, and the nuts or seeds inside are packed with even more goodness.

Flying foxes are the world's biggest bats. They eat only fruit and their favourite is ripe figs. They often fly over 50 kilometres to find them. In the rainforests of south-east Asia, fruiting fig trees attract hornbills, gibbons, giant squirrels and fairy bluebirds. There is some squabbling, but no serious fighting. They are all too busy enjoying the feast.

A red-knobbed hornbill peels a fig with the tip of its massive beak before tossing it in the air and swallowing it in a single gulp.

Nutcrackers

Some nuts and seeds grow in such a way that only animals with the right body parts can reach them. The agouti looks a bit like a giant tailless rat. Its front teeth are so sharp that it can crack open a hard Brazil nut shell to reach the flesh inside.

Brightly coloured macaws are expert nutcrackers too. They can crack and eat all kinds of seed, including poisonous ones. After swallowing seeds that would kill many animals, macaws fly down to the riverbank and scoop up beakfuls of clay. This mud pudding works as an **antidote** to protect the birds from the poison in the seeds.

As well as eating Brazil nuts, the agouti also feeds on fruit, seeds, leaves, buds, and even fungus.

Insect collectors

As they feed, leaf-eating and **nectar**-sipping insects are being hunted. For many animals they make perfect bite-sized snacks, full of **protein**. Catching live insects is not as simple as eating a leaf or drinking from a flower because the insects can run, hide or fly away. Insect eaters are called **insectivores**. They are experts at finding and catching their food.

The giant anteater from South America has a tongue over half a metre long that can flick out 150 times a minute. It eats 30,000 termites a day.

The insect-eating chameleon stays hidden until an insect lands near by. Then it catches the insect by shooting out its long, sticky tongue at high speed. Most spiders use sticky silk to trap insects, but not all spiders build a web. The female bolas spider uses one long thread of silk, weighted down at one end with a drop of glue. The spider swings this sticky thread at any insects that fly within range. When an insect flies into the thread, it sticks in the glue.

A chameleon's tongue shoots out so quickly that only a high-speed photograph can freeze the action.

Flying insectivores

A swift uses its gaping-wide mouth like a funnel to collect as many insects as possible. It may catch 10,000 insects in a single day. At night, insect-eating bats take over. Bats fly in total darkness and can track down a meal that they can not even see. Bats navigate by echo-location. They produce high-pitched squeaks up to 200 times per second. The sound bounces back off any nearby objects and the bat's huge ears pick up the echoes. A bat's hearing is so sharp that it can find the exact position of a flying insect in this way.

Helping each other

Some animals rely on others to help reach their food. The honeyguide eats bee grubs and wax, but it needs help to reach them. It relies on the ratel, or honey badger, to open up bees' nests. The honeyguide leads the ratel straight to a bees' nest. The ratel stuns the bees with a foul-smelling scent and then digs out the honeycomb. When it has finished eating, the honeyguide feeds on the leftovers.

Cleaners

African oxpeckers use giraffes and zebras as walking bird tables. The birds hitch a ride while they peck out and eat the blood-sucking ticks on the animals' skin. The remora, or suckerfish, behaves like an underwater oxpecker. It clings to sharks, sea turtles and even whales, using a powerful sucker on its head. It feeds on the scraps that they leave behind at meal times.

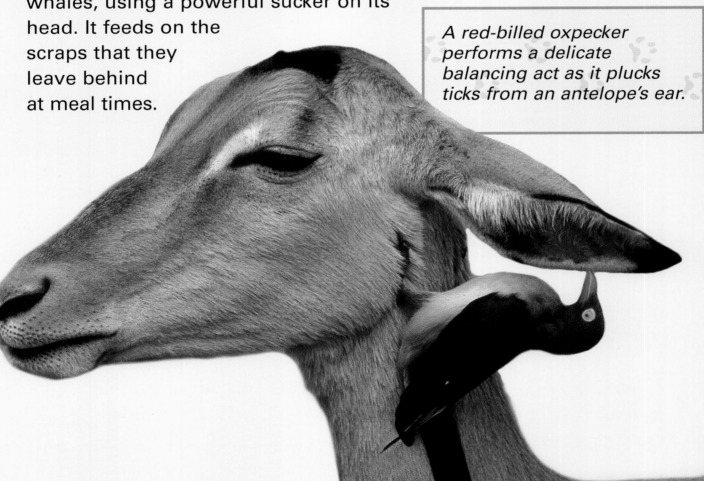

A red-billed oxpecker performs a delicate balancing act as it plucks ticks from an antelope's ear.

The cleaner wrasse has even stranger feeding habits. Bigger fish queue up at special cleaning stations. They are waiting to be cleaned inside and out. The wrasse picks off all the tiny fish-lice. It swims in and out of their mouths and cleans between their teeth. It will even eats bits of dead skin and fungus.

Green tree ants in Australia eat every caterpillar that they find, except one kind, the caterpillar of the common oak blue butterfly. This caterpillar squirts out a sugary drink and the ants 'milk' it. In return, they build it a shelter and act as bodyguards, driving away its enemies. When animals use each other in this way, scientists call it **mutualism**.

A large fish hangs in the water, keeping its body still while cleaner wrasses go to work on its scales.

Blood suckers

Sometimes the relationship between different animals is one-sided. One animal takes advantage of another and gives nothing in return. Creatures that feed on other living animals are called **parasites**. Their victims are known as **hosts**.

A parasitic wasp feeds its grubs on living flesh. It digs out an underground store and fills it with caterpillars and spiders. These have been **paralysed** by the wasp's sting. The wasp then lays an egg on each one. Then it seals the entrance to the store. When the grubs hatch, they eat their food supply alive.

Ticks are blood-sucking relatives of spiders. They bite through animal skin and use their tiny snouts like drinking straws to suck up the blood.

A female potter wasp stuffs a paralysed caterpillar into the mud jar that she has built to protect her egg.

The vampire bat also feeds on blood. It uses its nose to sniff out the best drinking spots where the animal's blood vessels are closest to the skin. It makes a cut with its razor-sharp teeth. As the bat drinks, a special **anti-coagulant** in its own **saliva** keeps the victim's blood flowing.

Rafflesia flowers take all the food they need from the host vine on which they grow.

Plant parasites

The largest flower in the world is called rafflesia. It is found deep in the jungles of Borneo. For most of its life, the plant stays hidden. It grows inside one particular kind of vine. When rafflesia is ready to flower, lumps begin to appear on the roots of the vine. These lumps are flower buds, which are growing under the surface. When they finally burst out, each flower has five petals the size of dinner plates. A rafflesia flower measures over a metre across when fully open and can weigh up to 24 kilograms.

Big appetites

Some animals have to eat almost non-stop to get enough energy just to survive. The world's busiest eater is probably the sand grouse. This bird has to eat up to 80,000 seeds a day. A grazing hippopotamus uses its massive mouth like a lawnmower and swallows 60 kilograms of grass in a single night.

Many of the world's biggest eaters feed on very small creatures. Humpback whales feed on krill. These are tiny shrimps that gather together in huge shoals. Whales can swallow thousands of krill in a single gulp.

Humpback whales drive krill into one big ball of food and then swim up underneath the shoal with their mouths gaping wide.

As the egg-eating snake pushes an egg down its throat, its sharp backbone works like a saw, cutting open the shell.

Big mouth

A big mouth is also useful for swallowing large items. The green imperial pigeon can swallow nutmegs bigger than its own head. It does this by unhooking its beak and opening wide.

The African egg-eating snake has even bigger ideas. This snake will steal eggs from a bird's nest. It can swallow an egg whole by **dislocating** its jaws, stretching its throat and slowly pushing the massive bulge down into its stomach. Just when it looks like the greedy snake will choke on its own meal, the egg cracks as it scrapes against the snake's spiky backbone. The snake swallows the yolk, and spits out the empty shell.

Predators

Predators are animals that hunt and kill other animals for food. Those predators that eat meat are called **carnivores**. Predators hunt animals that can run, fly, hide or even fight back. To catch their **prey**, predators need to be strong, fast, clever or very patient.

A killer whale launches itself onto the beach and grabs the nearest sea-lion pup in its powerful jaws.

Killer whales travel in groups called schools. They gang up on much bigger whales, often a female and her calf. They take it in turns to attack. They bite chunks out of their prey until it is too exhausted to fight back. On the coast of South America, killer whales even swim up on to the beaches to catch sea-lions. They wait for a big wave to carry them close to the sea-lions and then strike.

In late summer, grizzly bears go fishing. Every year, millions of salmon leave the sea and swim up the rivers, back to where they were born. The grizzlies are waiting. The fish battle against the **current** and leap up waterfalls. The bears wade into the river and try to catch them. They then pin down the wriggling fish with their needle-sharp claws. They rip off the skin, munch the tastiest bits of flesh, and throw away the **carcass**.

All these large **predators** have something in common: they are at the top of the **food web**. In other words, no other animal is big enough, or fierce enough, to try to eat them. At least, not while they are still alive.

The powerful jaguar is one of the deadliest predators in the jungles of Central and South America.

Nature's vacuum cleaners

All plants and animals end up as food for something else. A fierce **predator** does not live forever. When it dies, animals called **scavengers** will eat its rotting flesh. Every giant tree that falls in the rainforest will disappear completely one day. It will be eaten away by fungus, beetle grubs and other small creatures.

On the African plains, every dead and dying animal attracts a hungry crowd of scavengers. Cackling hyenas tear at the **carcass**. They are like walking dustbins. Their jaws are strong enough to crush the bones of any animal. Vultures and marabou storks often squabble over the last scraps of meat. Squirming maggots also feed on the rotting flesh. Anything left over rots into the soil and turns into plant food. Nothing goes to waste.

A flock of white-backed vultures gathers around the carcass of a large antelope.

Dung beetles collect fresh animal droppings. They roll them into steaming food balls before pushing them down into an underground storehouse.

Load of old rot

Some scavengers prefer to eat dead leaves or rotting wood. In South America there are giant millipedes that grow up to 20 centimetres long. These millipedes crawl through the **leaf litter**, and chew up the rotting vegetation. Even the tiniest animals play an important part in recycling waste. Termites eat rotten wood and dead leaves and vacuum up all the 'rubbish'. They re-use it by turning it into food. A colony of termites has a huge appetite. It will eat more than all the other animals in the neighbourhood put together.

Fact file

Plants and animals are linked by what they eat and what eats them. Scientists call this a food web. The food web below shows how some of the plants and animals in the Borneo rainforest depend on each other for food.

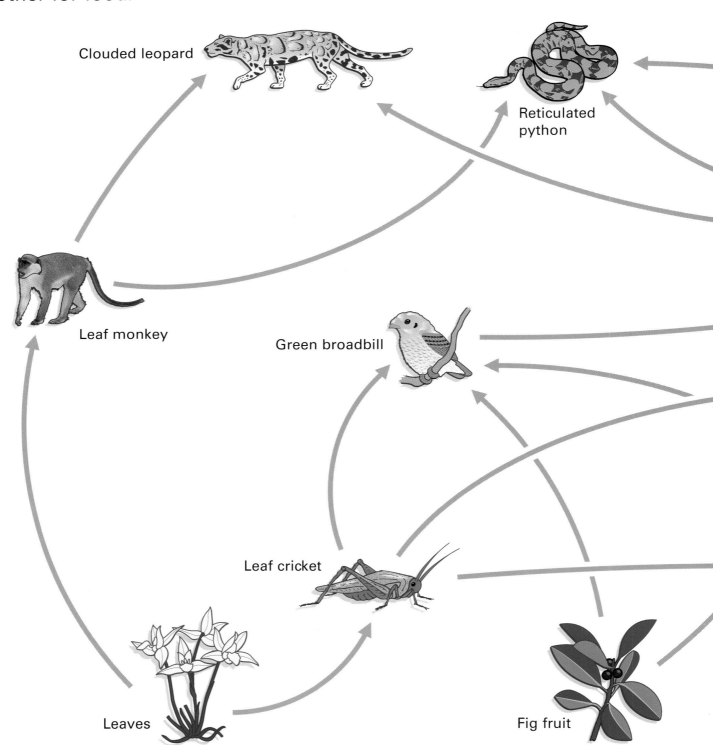

How to read the food web

Each arrow points from the animal or plant that is eaten to the animal that eats it. For example:

Termite → Pangolin → Python

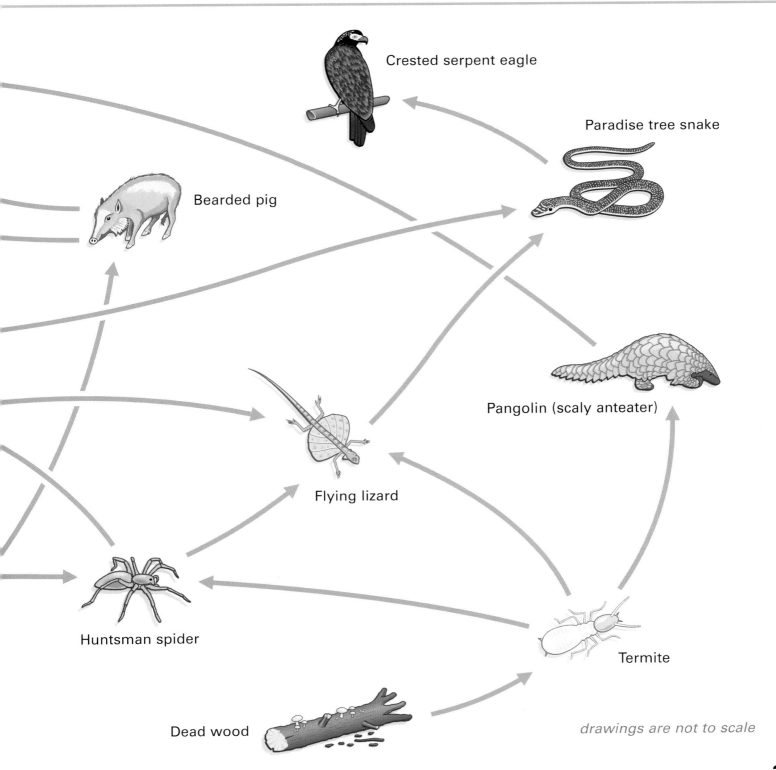

drawings are not to scale

Glossary

acidic containing strong chemicals that dissolve solid food

anti-coagulant substance that helps to keep the blood flowing around an animal's body

antidote cure for poison

bacteria tiny organisms in the soil, water and air. Some kinds can bring about decay in dead plants and animals

carcass body of a dead animal

carnivore animal that eats meat

cellulose main part of a plant's cell walls

current flowing water or air

dislocating to put out of proper place

dissolve to turn from solid to liquid

epiphyte a plant which grows on another one for support

food web relationship between plants and animals that feed on each other

graze to eat grass

host plant or animal used by a parasite to help it feed or grow

insectivore an animal or plant that eats insects

leaf litter layer of dead, rotting leaves covering the forest floor

mineral substance found in rainwater, soil and food

mutualism partnership between different plants or animals in which they depend equally on each other

nectar sugary liquid produced by flowers

nutrient the parts of food which living things need for energy or to build new cells

omnivore animal that eats both plants and other animals

paralysed alive, but unable to move

parasite plant or animal that feeds on another living plant or animal

pest animal or disease that destroys crops and other valuable plants

photosynthesis technique used by plants to turn sunlight into energy

predator animal that hunts and kills other animals for food

prey animal that is killed and eaten by predators

protein food that helps animals and plants to grow

reproduce to produce offspring

ruminant animal that has to chew its food more than once

saliva a liquid that is found in the mouth and helps to break up food

sap juice inside plant leaves and stems

scavenger animal that eats leftovers, such as dead meat

starch kind of food stored in plants

Index

agouti 15
anteaters 5, 16
ants 8, 13, 19

bats 14, 17, 21
bears 4, 25
bees 18
blood-sucking animals 20–1

carnivores 24
caterpillars 4, 9, 10, 19, 20
chameleons 17

digestion 9
dung beetles 27

epiphytes 6

fish 18–19, 25
food web 25, 28
fruit, nuts and seeds 14–15

giraffes 11
grazing 10

hippopotamuses 22
hornbills 14
hummingbirds 12
hyenas 26

insectivores 7, 16–17

locusts 9

macaws 15
millipedes 27
monkeys 9, 10
mutualism 19

nectar 7, 12, 13

okapi 11
omnivores 4
oxpeckers 18

parasites 20–1
photosynthesis 6
pitcher plants 7
plants 6–11, 21
poisonous plants and seeds 11, 15
predators 24–5, 26

rafflesia 21
ruminants 9

scavengers 26–7
snakes 5, 23
spiders 5, 17, 20
swifts 17

termites 27
ticks 18, 20

Venus fly-trap 7
vultures 26

wasps 20
whales 22, 24

Titles in the *Amazing Nature* series include:

Dramatic Displays — Hardback	0 431 16652 8
Fantastic Feeders — Hardback	0 431 16650 1
Ferocious Fighters — Hardback	0 431 16651 X
Incredible Life Cycles — Hardback	0 431 16662 5
Magnificent Movers — Hardback	0 431 16660 9
Marvellous Migrators — Hardback	0 431 16653 6
Powerful Predators — Hardback	0 431 16661 7
Super Survivors — Hardback	0 431 16663 3

Find out about the other titles in this series on our website www.heinemann.co.uk/library